Telecom Technologies Simplified for everyone:
Evolution of Technologies and Services

Pankaj Gupta
(ECE, MBA)

Copyright © May. 2022, Pankaj Gupta,

Updated Feb.2024 (Combined with "Simplifying Cloud Computing for everyone")

All rights reserved.

Dedication

This book is dedicated to the everyone who wants to learn basics of telecommunication technologies and associated services in a simplified manner and eager to know that how the telecommunication services are evolved during the last few decades.

In addition, I have added a section on Cloud Computing for the people who want to understand the concept of cloud services and business model. It simplifies the complexity of cloud for everyone who just need to understand the basics to enhance their knowledge.

Although the book is copyright of the author (except the references), but it can be used for training purposes for the benefit of everyone to gain the conceptual understanding of technology and services.

How is this book written?

How is this idea born? To explain, let me introduce myself.

I am a communication engineer by choice, by hobby, and by profession. This is because I chose this field for myself. And I have worked with various Telecom Companies and multinational corporations for 26 years. Most of the years I worked in network operation, customer support and handled customer escalations as well.

I used to work round the clock. You know, the operation people are available 24x7, 7 days a week and 12 months in a year. That is how you can use the telecommunication services uninterrupted, day and night, in hot summer and chilly winters, whole year and always available in need for you.

Have you ever observed the services which are always available to you? May it be the telephone exchange, which is always on, or any transport service viz. rail, road, or aeroplane. All these services are run round the clock without interruption. That is why the operation people must work round the clock to make this happen.

So, for telecom world, I was one of them. I was the one who handled Network Operation where I had to ensure that all the Telecom

Equipment work without fail. And if anything happens, in case of any fault occurs that had to be rectified as earliest as possible. Because that telecom network was the network in core, it was the backbone of all the telecom services. So, if the network is running smoothly, your services will run smooth. May it be land line telephone service, cellular mobile service, internet service or any other data communication Service.

In service operations, whenever any problem occurred in customers links, they used to call my helpdesk and I used to attend their call and start fault diagnosis. For initial few years, I worked day and night as one-man army in network operation and customer support. However, as the number of customers grew up, my organization had also expanded then functions were divided, and divisions were created. I got to handle planning and provisioning of the data communication network and services. It was then I started working in day hours and I was able to enjoy the sunshine on regular basis.

My company used to provide data communication services to software exporters for data transfer from India to rest of the world. So, during my tenure in network operation in 2000, a new team of engineers joined. Being an experienced guy, I was given the task of explaining to them about our network and the services. I took them for a tour in my International Gateway facility showing Satellite Earth Station, transmission equipment room and server room.

While explaining to them I got an idea why not to create a structured training module for new joiners, fresh engineers so that they will get an overview first, and then they will learn step by step, in a sequential way. Then I spent a few days to create my first training module that included classroom training as well as practical session on floor with the real-life activities.

And that was a time when I explored my skill of training, my passion for training and it became my additional skill. Since then, I continued that practice of training people in my further jobs with other organizations as well. So, at one point of time during my corporate career, I was transitioning from one company to another company, I got some time and thought why not to create my own training domain and share my learning with freshers to help them to **"Cross the bridge" from "Student Life to Industry"**.

Then I created the concept of "Cross the Bridge" and started making the framework. For some time, it remained pending due to other priorities of survival. Finally, I thought to write it now, and share in the form of a book so that I shall

How is this book written?

be able to share my experience, help you to learn the basics and guide you to cross the bridge.

But, reading a textbook is boring usually, specially, if it's intended for learning something to enhance your knowledge. Therefore, I have chosen the story format to explain telecom services from the user perspective first, relating it to my professional journey and then I dived a bit deep inside to give an overview of the underlying technologies without making it complex like engineering theorems.

Hope this story will help you to enhance your knowledge and understanding of the telecom services whether you are going to appear in any job interview in telecom industry or you are already in the industry with different profile, but you want to learn the basics of telecom technologies and services to get a high-level understanding.

Pankaj Gupta

Table of Contents

HOW IS THIS BOOK WRITTEN?	**5**
TABLE OF CONTENTS	**10**
MY INTRODUCTION WITH COMMUNICATION	**14**
1990S: BEGINNING OF A REVOLUTION	**17**
NEED OF COMMUNICATION TECHNOLOGY	17
HOW DOES A LANDLINE TELEPHONE WORK?	19
MULTI-ACCESS RURAL RADIO SYSTEM (MARR)	21
CELLULAR MOBILE	22
DATA COMMUNICATION	24
NEED OF DATA COMMUNICATION SERVICES	25
HOW DOES POINT-TO-POINT LEASED LINE WORK?	26
INTRODUCTION OF INTERNET	28
HOW DID INTERNET WORK?	30
2000: DECADE OF EVOLUTION	**31**
BROADBAND INTERNET	31
MULTI-PROTOCOL LABEL SWITCHING (MPLS)	33
VOICE OVER INTERNET PROTOCOL (VOIP)	36
EVOLUTION OF MOBILE TECHNOLOGIES	37
TRAI	38
2010: FURTHER EVOLUTION AND MATURATION	**40**
CONVERGENCE	40

SOFTWARE DEFINED WAN (SD-WAN)	42

FURTHER DISCUSSION ON TECHNOLOGIES — 44

GENERATIONS OF MOBILE COMMUNICATIONS	44
COMPARISON CHART OF MOBILE GENERATIONS	46
DIFFERENCE BETWEEN MOBILE GENERATIONS	47
POINT TO POINT LEASED CIRCUITS	47
INTERNATIONAL PRIVATE LEASED CIRCUIT (IPLC)	51
INTERNET	54
MULTI-PROTOCOL LABEL SWITCHING (MPLS)	59

SIMPLIFYING TECHNICAL TERMS — 62

VOICE COMMUNICATION	62
ANALOGUE AND DIGITAL COMMUNICATION	65
MODULATION	67
CELLULAR TECHNOLOGY	70
SATELLITE COMMUNICATION	72
DATA COMMUNICATION	75

FUN TIME — 81

AFTERWORD — 83

CLOUD SERVICES — 85

THE CONCEPT — 87

WHAT IS CLOUD?	87
CLOUD BUSINESS MODEL	88
INTERNET	89

Table of Contents

EXPLAINING CLOUD COMPUTING	90
ADVANTAGES OF CLOUD SERVICES	92
CLOUD COMPUTING SERVICES	**94**
INFRASTRUCTURE AS A SERVICE (IaaS)	94
PLATFORM AS A SERVICE (PaaS)	95
SOFTWARE AS A SERVICE (SaaS)	95
CLOUD DEPLOYMENT MODELS	**97**
PRIVATE CLOUD (ON-PREMISES)	97
COMMUNITY CLOUD	98
PUBLIC CLOUD	99
RELATED TERMS	**100**
DATA CENTRES	100
VIRTUALIZATION	101
INTERNET OF THINGS (IoT)	102
OFFICIAL DEFINITION OF CLOUD COMPUTING	103
FROM THE AUTHOR	**105**

Pankaj Gupta

Chapter-1

My introduction with communication

Way back in late 1980s when I decided to study electronics engineering, there was an option of choosing an elective out of three subjects and I chose communication engineering. Although other options were easier to score but I wanted to explore the communication world, more specifically telecommunication and I scored highest in my subject.

Subsequently in early 1990s when I got my first job offer from a telecommunications company in Hyderabad, I travelled around 1500 kilometres from my hometown in UP and joined that job. I was all alone far away from home, and I needed to connect with my home.

In those days only landline telephone was available, and it was also luxury that time as only rich people or VIPs owned the telephone. Being a middle-class family, we didn't have telephone at home. So, the only medium of communication with home was to write letters. But posting letters and waiting to reach them used to take a week or ten days. Further, getting reply took another week at minimum.

It was then when I realized the need of telephone at my home as I needed to be connected to my parents more frequently being in a touring job. So, I figured out a way to get telephone connection at my home. My grandfather was a freedom fighter, and he was entitled to get a telephone, so I applied for a telephone connection on his name.

Since then, I have been a witness of the journey of telecommunication services for three decades. I have seen evolution and phases of development from landline telephone to cellular mobile for voice communication, from point-to-point links to internet and MPLS for data communication.

My introduction with communication

~ 16 ~

Now, I am presenting here the gist of my telecom journey from being a user to becoming a technologist.

Pankaj Gupta

Chapter-2

1990s: Beginning of a Revolution

Need of communication technology

If I am speaking to you and you are listening to me or I am writing to you and you are reading it, both are examples of communication. When I am writing, this is being published in form of a book and it will go to market. You buy it from market, and you read it. So, I am communicating some information and you are receiving it.

But in case of voice communication when you are near to me, you can hear me. When I speak, sound waves originate from my mouth and travels in air and reaches to your ear which is acting as a receiver. Then your mind decodes the message and interpret the information. Then you say that you can hear me and understand the message.

However, these sound waves from mouth can travel very short distance and only people near me can hear my words. Even if I shout, people a little far, maybe up to 100 meter can hear me. But what happens when you want to communicate to a person very far in different part of the city or in the different city, or the different state, or all together in the different country.

So, we needed a medium or carrier to transport our voice for long distance.

It's similar to our journey in real life. When you travel long distance, you need a carrier or a transport system and there are different means of transport available for us viz. road, railways, and air. You can sit in the bus, and you can be transported to different place. Alternatively, you can sit in the train and travel to different place or you board in the aeroplane and fly to another place. So, you see there are different modes of transport system.

But what is the difference in these transport systems?

The speed of Transport is different for bus, train, and aeroplane. And how they travel in the geographical area is also different. Like bus is travelling on the road, train is traveling on the rail lines and aeroplane is traveling through air. But finally, they are making you to reach your destination. Their ultimate objective is to transport you to your destination.

Similarly, when you want to talk to someone on telephone, that means you want

to send your voice to the person sitting somewhere at remote place. Hence, you need a transport system or a carrier which can carry or transmit the sound waves originated from your mouth and take it to the ear of the other person, the receiver.

Here also, there can be different mediums or mode of transport to carry the sound waves. Initially, landline telephone was one of the medium, invented by Graham Bell in 1876.

How does a landline telephone work?

It's very easy from user perspective; you lift a phone receiver, dial the number, other side person picks up the phone receiver, and you start talking and two-way communication is completed. But what happens technically?

How are we able to transmit our voice far away from one place to another place?

The deeper I thought, deeper I dived and learned more and more about technology. Landline telephone network was the only medium of transport to carry our voice in initial days. Refer the diagram below.

1990s: Beginning of a Revolution

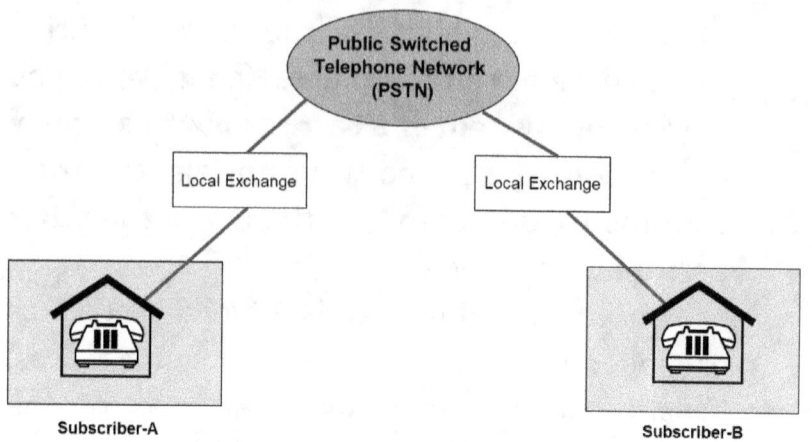

Fig.1 A simple representation of landline telephone system

There is a telephone receiver at our home which is connected to the nearest telephone exchange through physical line of copper wire. And that copper wire transports our voice and signalling information to the nearest telephone exchange. When we lift the handset, we hear a dial tone that is an indication that we can dial the number of the recipient. When the number is dialled, this signal travels to the local exchange. And that telephone exchange is further connected to the main exchange, may be through copper wire, or through the fibre or through high frequency microwave link.

This telephone system is called Public Switched Telephone Network, abbreviated as

PSTN. Until early 1990s, it was an age of landline telephone in India and Department of Telecommunication (DOT) was the only service provider (later, DOT has created separate entity to provide the Telecom services and named it Bharat Sanchar Nigam Limited around 2000). So, Government owned the telephone services and there were very few subscribers. We were totally dependent on the lineman of the telephone exchange in case any fault occurred to our phone lines.

Multi-Access Rural Radio System (MARR)

The company I joined in Hyderabad used to make multi-access rural radio system abbreviated as MARR. In early 1990s, the reachability of landline was mostly limited to developed cities only. Copper cable laying was not feasible in many of the rural areas, remote locations, or the area which are difficult to reach. Then this MARR system was evolved to extend landline telephone services to such areas.

In MARR system, there used to be one base station and multiple remote stations. Base Station unit (BSU) acted like a small exchange, and it was connected to nearest telephone exchange on the back end. And Remote Stations Units (RSU) were placed at gram panchayat or a central place in the villages. Then a telephone instrument was connected to the remote station unit. So that way, we used to provide telephone connectivity to rural areas and in difficult terrain.

Landline telephone was the only medium of communicating until 1995 when the cellular mobile service was introduced in India. And then MARR system became obsolete with the entry of the cellular technology.

Cellular Mobile

You all must be using mobile phones by now. But do you know how does this cellular mobile system work?

We have got a mobile phone in our hand. When we speak, it takes our sound waves as voice signal, convert it into radio waves, and transmit it to the nearest mobile tower called Base Station Transceiver system abbreviated as BTS. Now, BTS transmit the signal to BSC (Base Station Controller) which interfaces with Mobile Switching Centre (MSC). One MSC can serve to several BTS, so when you are travelling your call is hand off from one BTS to another BTS and your talks goes on uninterrupted.

Pankaj Gupta

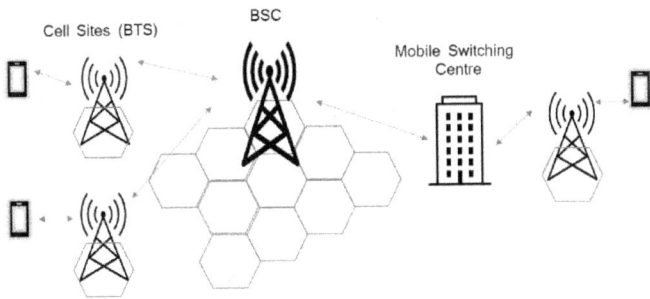

Fig.2 Cellular Mobile System

If the subscriber, you are calling is present in the coverage area of the same BTS then call is forwarded to that person directly. But if the subscriber is in some other location in the coverage area of another BTS, then call goes to MSC and there it is forwarded to the called subscriber. So, this is how Cellular Mobile communication work. Technically, it's called Global System for Mobile communication abbreviated as GSM.

To provide seamless connectivity from any cellular mobile subscriber to any landline telephone subscriber, mobile network was interconnected with landline telephone network as well, so that any mobile subscriber could call land line telephone subscriber and vice versa.

Do you remember when did you use cellular mobile phone first time? Though the cellular mobile phone was introduced in 1995 but took a few years to reach it to common people. In

1990s: Beginning of a Revolution

initial days mobile handsets were very costly and mobile call charges were also very high. We were even charged for incoming calls as well and that too was around 16 rupees per minute. I have used cellular mobile phone first time in 1997 when I was working with a "Data Communication service provider".

Now let me explain you this new term "Data Communication".

Data Communication

"Transfer of information in electronic form from one computer to another computer in the same office campus or from one office to another office located far away using a communication medium or transport system is usually termed as data communication".

My company used to provide high-speed **internet** connection and **point to point leased line** to the software companies. Since the other end of the customer was abroad, then point to point link becomes international and called IPLC (International Private Leased Circuit). It was for the companies who wanted to export their software from India to abroad as my company was formed to promote the software export from India. While Videsh Sanchar Nigam Limited (VSNL, a government entity) was open to all industries to provide the same services i.e., Internet and IPLC. And it was their monopoly until 2000 when telecom sector was open for private players. (VSNL was also privatized later and it's Tata Communications now).

Therefore, in parallel to evolution of cellular mobile services, late 1990s was also the era of beginning of Data Communication services in India.

Need of Data Communication Services

Now, let's understand why we needed data communication services. If we sit on time machine and travel little bit back on the scale of time and observe how did communication happen between companies or multiple offices of a company earlier then we recall that if they needed to send any information, then they had to write letters by hand using pen and paper or type and print on papers and send it to other person or their branch office or head office through postal department. That could be a normal post or registered post, or they could use the services of a private courier. It was very slow, and time taken process as one needed to take it to post office and then it was transported from one city office to another city office. It also needed more resources in terms of manpower and public transport systems.

With the evolution of **point-to-point leased line**, it became easy for them to transfer data from one computer in office to another computer in another office using the point-to-point link. So, whatever data they want to send could be transferred via this link and it facilitated an easier, faster, and efficient way for corporate offices to communicate between each other.

How does Point-to-Point Leased Line work?

The figure below shows a typical representation of a point-to-point link between two of the customer's offices. Typically, these modems provided bandwidth from 64 Kbps to 2 Mbps. Higher speed modems are also available now a days.

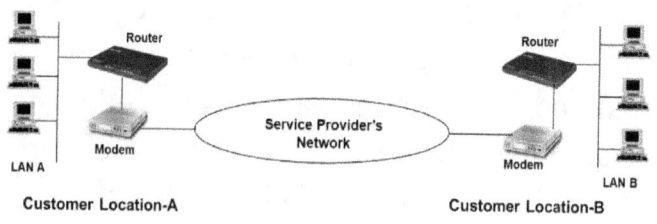

Fig.3 Point to Point link Connectivity Diagram

To transfer their data using the link, customers connect their end equipment (usually routers) to the modem installed by the service provider. There is no router used in the service provider's network. It is a dedicated communication channel given to customers between two of their office locations.

In case of International Private Leased Circuit (IPLC), other end the customer location would be somewhere abroad and there will be another service provider involved to provision the foreign half circuit. Indian half circuit is maintained by Indian provider and these two providers work in collaboration to provide services to the prospective customers.

Limitation of Point-to-Point Leased Line

When multiple offices had to be interconnected, then it was difficult to connect via point-to-point links because each office had to be connected to all other offices using point-to-point links. Then it used to create a mesh which proved very costly and complex system to manage. Then a new service evolved i.e., Internet which addressed this issue and simplified the connectivity.

1990s: Beginning of a Revolution

Introduction of Internet

Around mid-90s with the introduction of internet in India we started providing internet services as well. **Internet** addressed the problem of point to multi-point connectivity amongst multiple offices. Now, every office could connect to the internet and then it could communicate with any number of other office locations without any restrictions.

Internet provided a seamless communication medium from any part of the globe to any other part of the globe, so it became an easier and cost-effective way for data transfer for the corporate companies as well as individual users.

Internet became a revolution. It had revolutionized the world in the ways we used to transfer data. It became very popular not only for data transfer but for many other applications as well.

Pankaj Gupta

Applications of Internet

Electronic mail service, file transfer, world wide web browsing, personal chat and so many other applications have been introduced after entry of internet in India. Individual users and business houses started using internet for their data transfer via email and for file transfer. Following trends were observed thereafter:

- **Email** started replacing letters which used to be sent via postal department.
- Information available on **world wide web** helped students and everyone to access whatever information they needed on immediate basis for which they used to visit libraries earlier and search for relevant books on their topics.
- Corporates got an electronic medium in form of **websites to advertise their services** to the world in an easy and cost-effective manner in comparison to earlier advertising practices of manual promotions and paper print advertising.
- A lonely person was able to **find a friend** across the globe with matching interests in chat groups.
- **Professionals** could discuss their issues on their subject in **discussion forums** online.

1990s: Beginning of a Revolution

How did Internet work?

Initially, internet was accessible via dial up modem in in early 1990s and speed was limited to 56 Kbps only. Whenever we needed to use internet, we had to dial the given telephone number and got connected to internet. And we were billed as per time duration between connection established and disconnection. So, this was on demand service and billing was as per usage time. And Videsh Sanchar Nigam Limited (VSNL) used to provide that dial-up internet services in initial days.

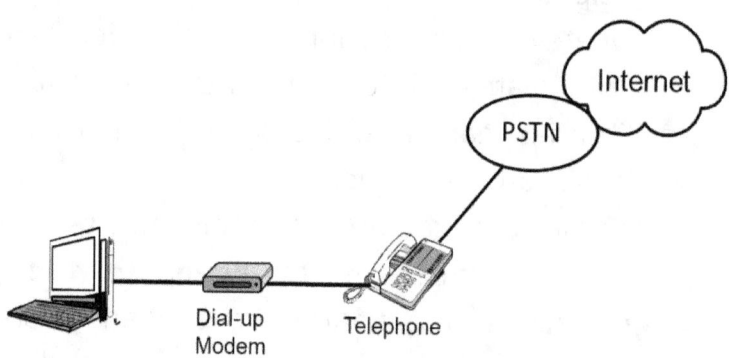

Fig-4 Internet access via dial up modem

Chapter-3

2000: Decade of Evolution

Broadband Internet

Very soon after dialup, high speed internet services became available with the advent of new technology called DSL (Digital Subscriber Line). DSL technology was able to provide **always on high-speed internet connection** using the same copper line which was used to provide telephone service at home.

Initially Internet connection having minimum speed of 256 kbps was termed as Broadband, but it has been changing with the advancement of new technologies. Telecom Regulatory Authority of India (TRAI) has proposed that a minimum of 2 Mbps speed is required for an internet connection to be called 'broadband' now.

Now in place of dial-up model, a DSL modem was placed at subscriber's end (customer premises), which was connected to the nearest exchange through existing copper wire used for telephone. An aggregator called DSLAM (Digital Subscriber Line Access Multiplexer) was placed at the exchange side which aggregated the traffic from multiple customers. This way always on internet was made available to the customers.

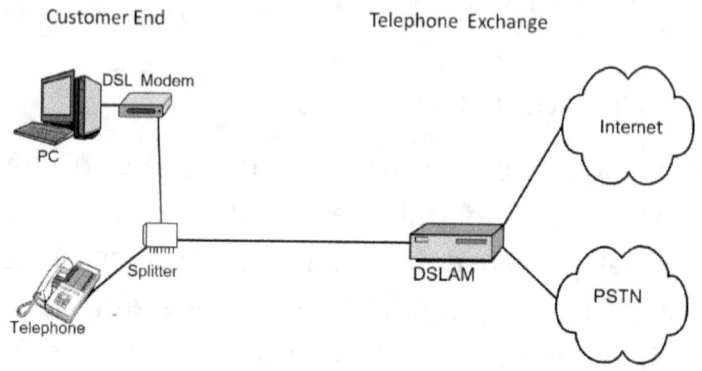

Fig.5 Broadband Internet using DSL.

This connection was also termed as internet leased line as the link from customer end to exchange is dedicated for that customer only. However, customer data traffic always travel through shared network after landing into public internet network.

Limitations of Internet

Internet created a revolution; it revolutionized the world and changed the way we live our life. However, internet is a public network. It's like a public road. You don't have any control on your road traffic when you come out from your home. When you drive your car or any other vehicle on the road you don't know how much time it will take to reach from one point to other point or from your home to office. It depends on the instantaneous traffic on the road you travel during driving.

Similar phenomena happen on internet since millions of users are browsing internet at any point of time being it a world-wide Network. Data traffic from millions of users are transmitted and received at a time. And there is no guarantee of speed that how much speed you will get and how much time your data will take to reach your destination or to your other office or any recipient. Further evolution in technology addressed this problem.

Multi-Protocol Label Switching (MPLS)

A new technology emerged in early 2000 named MPLS which offered a private network between corporate offices addressing the problems of speed, privacy, and security. MPLS allowed to create a secure and private Network for inter-office communication as it could connect multiple offices of a corporate in a cost-effective manner in comparison to point to point links and internet.

In internet any data packet sent on the internet network is analysed at each router and routing decision is made, then only data packets are transmitted to the next router. Therefore, checking, rechecking, and rerouting caused delay in packet transmission.

MPLS offered label switching where a packet is labelled to reach on the next hop only. Then in every hop, every router in the network just swaps the label, that is, remove the previous label, put the next hop label, and send it to the next router in the network. Here a Label Switch Path (LSP) is created to send a packet which predefine the route of data packet from source to destination.

How does MPLS work?

When the packet is transmitted from source to destination, first of all, from the customer edge router (CE) it reaches the first router on the MPLS network, that is called Level Edge Router (LER) or Provider Edge (PE) router. Only this router analyses the Packet and checks where is the packet to be delivered. And as per the destination it creates a label switch path, adds a label to that packet and sends to the next router in the network which is called providers router (P) which acts as Label Switch Router (LSR).

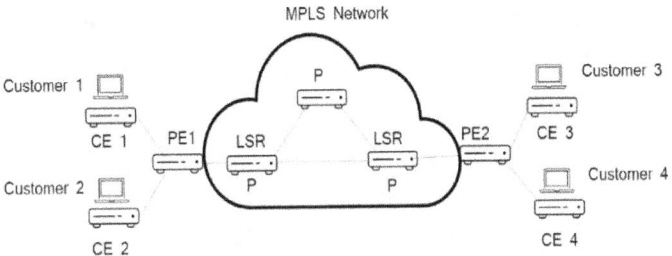

Fig.6 A simple representation of MPLS services.

In this way a packet keeps on traveling hop to hop, by just label swapping. None of the routers on the way analyses the packet and makes the routing decision. Only last router in the network analyses the packet and checks where it is to be delivered and then forwards it to the destination, the intended customer.

This saves the time and packet reaches on the destination router faster than internet. **That is why MPLS provided faster, and secure network for data transfer.**

To get MPLS connection, each of the customer's office locations had to be physically connected to the nearest point of presence (POP) of the MPLS service provider. All the customer locations who want to communicate with each other need to be connected to the nearest MPLS POP. And then virtual private network (VPN) is created over physical MPLS network. It's like a tunnel, which is created between office locations. So, any office can transmit data to any other office in a secure way and without any delay, which is likely to happen on the public network like Internet.

Voice over Internet Protocol (VOIP)

IP-based networks can provide real-time services such as voice and video telephony as well as non-real-time services viz email, browsing and other applications. When voice is carried over an IP network it is called Voice over IP. And if the IP network in this case is the public Internet, then it becomes Internet telephony.

How it is different than normal telephony?
Public Switched Telephone Network (PSTN) has been the primary means for voice communication. The PSTN is a connection-oriented network in which a dedicated channel is established for the duration of a communication.

IP Telephony provides alternative means of originating, transmitting, and terminating voice and data transmissions. VoIP applications can run over a range of devices, offering flexibility towards seamless communications. With advancement in new technologies, Internet telephony, or VOIP is becoming the primary medium for converging voice and data services.

Pankaj Gupta

Benefits of VOIP

As most of the corporates and consumers use internet, so we already possess the connectivity to the Internet. VoIP makes use of existing network and enable users to make real time voice calls, transmitted over the Internet (rather than traditional telephone networks). This enables service providers, and consumers to make savings by reducing the underlying costs of a telephone call.

Evolution of mobile technologies

Initially cellular mobile was created for voice communication only. Then further short message service (SMS) was added with development of Global Packet Radio Service (GPRS) technology. In further developments during late 2000, with the availability of broadband spectrum for mobile communication, high-speed internet service became possible on mobile itself. We were able to use mobile data service for internet browsing, video calls and live video streaming as well.

Phases of evolution in mobile technologies were termed as generations. Mobile with voice communication was called **1st generation** (1G). Introduction of text service as SMS made it **2nd generation** (2G). Then broadband mobiles with voice, data and video capabilities were called **3rd generation** (3G).

In December 2008, India entered the 3G arena with the launch of 3G enabled Mobile and Data services by Government owned Mahanagar Telephone Nigam Ltd MTNL in Delhi and later in Mumbai. After MTNL another government entity Bharat Sanchar Nigam Ltd. provided services in rest of India. Further many of the private telecom operators have also upgraded their network and started providing 3G services.

With evolution of 3G, we could browse internet on our mobile and watch videos with reasonably good performance. That's when mobile data was termed as mobile broadband.

TRAI

The entry of private service providers brought with it the inevitable need for independent regulation. The Telecom Regulatory Authority of India (TRAI) was, thus, established with effect from 20th February 1997 by an Act of Parliament, called the Telecom Regulatory Authority of India Act, 1997, to regulate telecom services, including fixation/revision of tariffs for telecom services which were earlier vested in the Central Government.

Pankaj Gupta

TRAI's mission is to create and nurture conditions for growth of telecommunications in the country in a manner and at a pace which will enable India to play a leading role in emerging global information society.

One of the main objectives of TRAI is to provide a fair and transparent policy environment which promotes a level playing field and facilitates fair competition.

Ref.: Internet resources

Chapter-4

2010: Further Evolution and Maturation

Convergence

"Convergence means integration of multiple services in one network and providing to customer on a single last mile or using a single device".

Although the convergence was in discussion much before 2010 and few services were launched but it has been in development phases and seem to have matured now. Let's get a high-level understanding.

Earlier we used separate access medium or last mile for telephone, Internet, and video services. With the integration of networks, it became possible to provide many services on a single last mile. For example, we got telephone line using copper wire. Then we got internet using satellite or separate copper wire or fibre line. And we received television videos using satellite direct to home (DTH) service. Initially voice and data services were integrated, and we started getting both the services on the same last mile from the service providers.

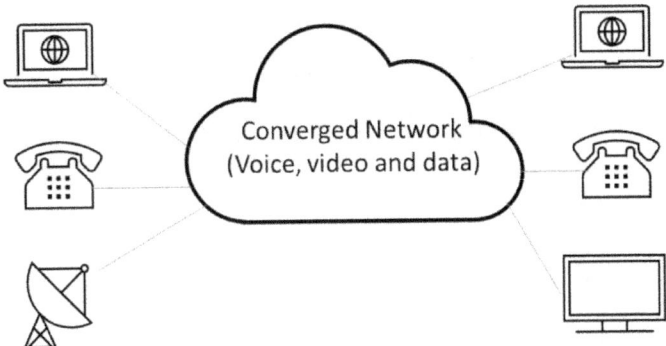

Fig.7 A simple representation of convergence

In further development, video service was also integrated making it a triple play. That means we could get telephone, internet, and television on the same last mile from the service provider. It was named IP TV initially launched by MTNL and then Airtel also provided this service for some time on the same copper list mile which was used for telephone services.

Now there are many service providers providing multiple services using same fibre last mile named FTTH (fibre to the home) or FTTB (fibre to the building). Telephone, Internet, Television, and video streaming, a bouquet of services is available on the same access medium (fibre) now.

It's a win-win equation for customers as well as service providers. Customers receive a single bill and a single point of contact for multiple services. And providers can sell multiple services

2010: Further Evolution and Maturation

over a single last mile reducing their cost incurred in network provisioning and operation.

Software Defined WAN (SD-WAN)

After MPLS, a new technology emerged; Software-defined wide area network. Before SD-WAN came in picture, if a corporate wants connectivity between multiple offices they had to get same kind of data communication service at all of their offices. Either they had to use Internet on all locations, or they had to take point-to-point or point to multipoint links. But after development of SD-WAN, customer has got the liberty to take any connectivity at any office and then, configure them in such a way that they look like they are only single WAN.

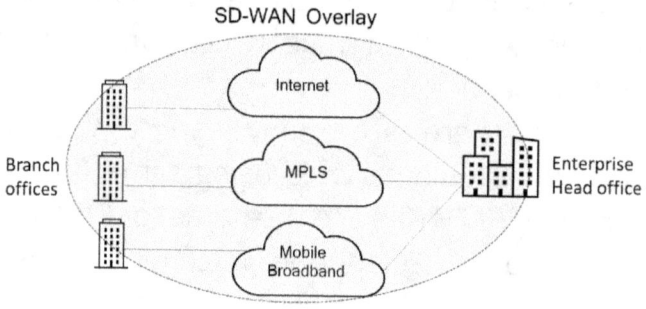

Fig.8 SD-WAN representation

How does SD-WAN work?

Point-to-point link used PPP protocol, internet used IP protocol and MPLS used MPLS protocol. But SD-WAN enables the network to create a virtual WAN which is an overlay network over existing network and configures a virtual WAN on the already available physical Network, irrespective of their type of connectivity using overlay protocol. It can be configured only on already available physical Network.

Now, let's dive a bit deeper to discuss more about technologies in further chapters.

Chapter-5

Further discussion on technologies

Generations of mobile communications

Cellular telephones began as two-way analogue communication using frequency modulation for voice and handover decision was taken by base stations in the first generation. It used Frequency Division Multiple Access (FDMA) in which each cell phone call is assigned an uplink and downlink frequency channel until the call is finished or handed off.

Second generation (2G) of mobile used the digital communication techniques and added new service viz Short Message Services (SMS) and Multimedia Message Services (MMS). It used Global System for Mobile (GSM) standards developed by European Telecommunications Standards Institute (ETSI) which used 25 Mhz frequency spectrum in 900 Mhz band.

Third generation (3G) technologies focussed on improving voice services, higher bandwidth and support of multimedia services which are developed by two technology standards: Universal Mobile Telecommunication System (UMTS) and CDMA-2000. The International Telecommunication Union (ITU) has defined 3G system which can support high speed data from 144 Kbps to more than 2 Mbps.

Fourth Generation (4G) of broadband cellular network topology is based on the capabilities defined by the ITU in IMT Advanced (International Mobile Telecommunication Advanced). According to ITU, a 4G network requires mobile device capable of carrying data at 100 Mbps. There are different network standards for 4G viz. Long-Term Evolution (LTE), LTE-Advanced, WiMAX and Ultra Mobile Broadband (UMB). In fact, the UMTS technology upgrade is termed as LTE to accomplish higher speeds along with lower packet latency.

Introduction of 4th generation of mobile technology made our life easier, and mobile has become our portable office.

5G is labelled as 'ultra-fast, ultra-reliable, ultra-high capacity transmitting at super low latency' by the National Infrastructure Commission in the report "5G Infrastructure Requirements in the UK" (2016). Facilities that might be seen with 5G technology include far better levels of connectivity and coverage. The term Worldwide Wireless Web or WWWW is being coined for this.

Further discussion on technologies

Telecommunication Service Providers (TSPs) are experimenting with transmission of signals on a whole new swath of spectrum of 20~50 GHz. This band, acknowledged as the millimetre (mm) Wave band, makes use of higher frequencies than the radio waves that have long been used for mobile phones.

Comparison chart of mobile generations

GENERATION	1G	2G	3G	4G	5G
DATA BANDWIDTH	---	14-64 Kbps	2Mbps	200Mbps	1Gbps
TECHNOLOGY	Analog cellular	Digital cellular	Broadband with CDMA, IP technology	Unified IP and seamless combination of broadband of LAN, WAN and WLAN	Unified IP and seamless combination of broadband, LAN, WAN, WLAN and WWWW
SERVICES	Mobile technology (voice)	Digital Voice, SMS,	Integrated high-quality audio and video	Dynamic information Access, Wearable devices	Dynamic information Access, Wearable devices with AI capabilities
MULTIPLEXING	FDMA	TDMA, CDMA	CDMA	CDMA	CDMA

Ref. Internet resources

Pankaj Gupta

Difference between mobile generations

Let's understand the difference between generations of mobile services at high level. From the user perspective, it added more and more features viz SMS (in 2G), Data Services, Internet Browsing, and video streaming (in 3G) during the phases of development. From the network perspective broader frequency bands are used with higher transmission power to reduce the latency and provide the faster service.

You can compare it with real life analogy of development of road network and highways. You have a small lane near your home and it's so narrow that you can only walk through it. In initial phase of development, it was widened, and you were able to drive two-wheeler only with limited speed. In next phase of development, it was widened to provide space to drive four-wheeler also now. Further development widened it more and multiple lanes are created to accommodate more and more traffic of vehicles and broad space for driving at higher speeds.

Point to Point Leased Circuits

In early 1990s Point-to-point link was the only communication medium to transfer data, between two of the offices when any company wanted to

Further discussion on technologies

send some information in electronic form from one office to another office located across the city, across the nation or anywhere across the globe.

When the two offices connected via point-to-point link are located in the same city then the service is called **"Local link or intra-city link"**. If the two offices are located in different cities within the same country, then this service is called **"National Long Distance (NLD)"** service. And when one end location of the customer is in India and other end is anywhere abroad then this point-to-point link is called **International Private Leased Circuit (IPLC)**.

See the concept is same here for point-to-point link, whether it's connecting two offices in a city or two offices within a country or two offices within a globe that is from anywhere in India to anywhere abroad. Hence, it is also called point-to-point leased line, and it is a clear Communication channel provided to customers to transfer their data securely, privately and in a faster manner between two of their office locations.

What is Local Loop?

Now, please understand that **Local Loop (LL)** term is used widely for the **last mile** connectivity from the customer premise to the nearest exchange or point of presence (POP) of the service provider. Any service, whether Internet, leased line or MPLS will need a local loop or last mile to extend the services to customer premises. It's called the last mile since it connects the last leg when you start from service provider's POP and walk towards customer. However, it's the first mile when you start from customer premise and walk towards service provider's POP.

Who provided the local loop?

Initially, during 1990s, Bharat Sanchar Nigam limited (BSNL) used to provide local loop or last mile connectivity and Videsh Sanchar Nigam Limited (VSNL, now Tata Communications) used to provide National Long-Distance Service (NLD) between two of the customer locations in India. After 2000, with privatization of telecom sector, many service providers got the license for Local Loop, NLD and IPLC and started providing these services.

What are the applications of point-to-point leased services?

These are exclusive, point to point links, which can provide high-speed transfer for data, voice, and video on the same link in a private and secure manner. It's a communication channel dedicated for that customer only and no mixing with any other customer's traffic.

Further discussion on technologies

Who is the target Audience?

Point to Point Link or Point to Point Leased Circuit is an ideal solution for applications which are time and/or content sensitive, as well as for establishing an integrated network to handle a variety of functions within an organization's wide area network. It is very useful for mission critical applications where the scope of error and delays in data transfer is extremely critical.

Is it different than other services?

Local Link or National Long Distance Leased Circuit is a dedicated point-to-point connectivity at subscribed speed between customer's two locations with no connectivity to the PSTN at either ends. Since bandwidth is totally dedicated to customer, these circuits provide secure, reliable, and high-speed connectivity.

How is the connectivity for National Long-Distance Link?

There are two components in the NLD Leased Circuit. One is the connectivity between the NLD POPs on the providers NLD network and other is the Local Loop connectivity between the NLD POP and customer premises. The figure below shows a typical connectivity for National Long-Distance link which connect two of the customer's locations within a country.

Pankaj Gupta

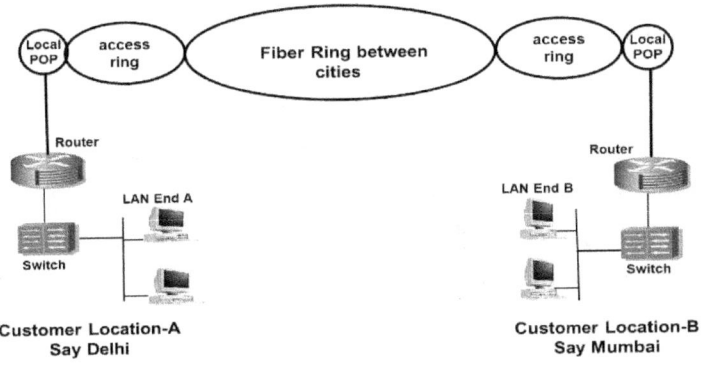

Fig. 9 A Typical connectivity diagram for NLD

International Private Leased Circuit (IPLC)

What is IPLC?

IPLC is a Point-to-Point link between two International Locations where one location can be India and the other location is anywhere abroad.

What are the applications?

When a corporate customer wants to send Voice, Data & Video traffic to their counterpart in abroad location in a secure and private manner, IPLC is an ideal solution.

Who is the target Audience?

Call Centers, Internet Service Providers (ISP), Software Development Companies are the target market segment for IPLCs where they want dedicated, secure & private connectivity between

Further discussion on technologies

their two offices one in India and another in abroad.

How is it different from other services?

Technically, IPLC is the same as any other point-to-point leased line where the subscribed bandwidth is reserved for the customer but geographically the other customer end is abroad in the case of IPLC.

How is the typical Connectivity Diagram of IPLC?

To provision an IPLC, Indian half circuit is provided by an Indian service provider but to provide the foreign half circuit, they need to collaborate any of the foreign service provider to complete end to end circuit.

Figure below shows the connectivity diagram for an IPLC. Initially during 1990s the international link used to be through satellite channels but after 2000 with the expansion of fibre networks, it is usually provisioned through submarine fibre cables now.

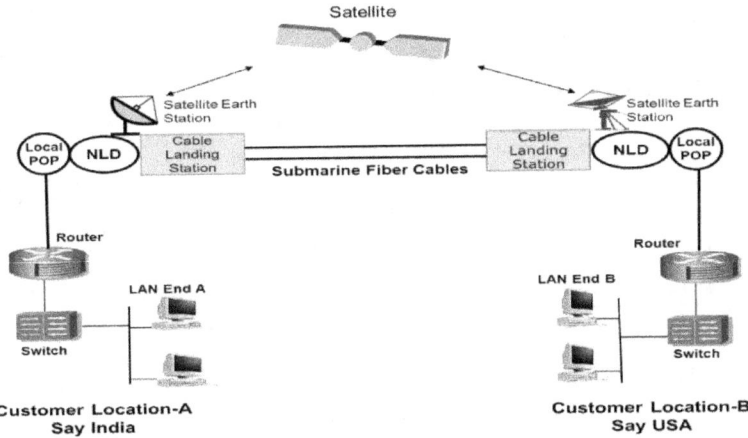

Fig. 10 Typical Connectivity for IPLC

Further discussion on technologies

Internet

What is the Internet?

The Internet is an electronic network of computers that includes almost every organization, university, government, and research facility in the world. Also included are many commercial sites.

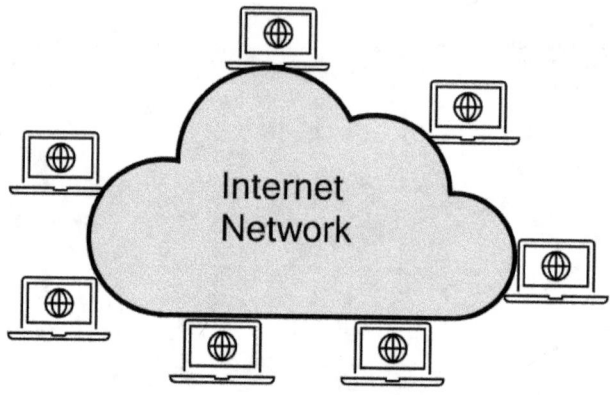

Fig.11 A representation of Internet

The Internet started with four interconnected computers in 1969 and was known as ARPAnet. There may be multiple ways to explain the Internet.

- A network of computer networks which operates worldwide using a common set of communications protocols.

- The vast collection of inter-connected networks across the world that all use the TCP/IP protocols.
- The Net," is a worldwide system of computer networks providing reliable and redundant connectivity between disparate computers and systems by using common transport and data protocols.

What are the applications of the Internet?
- Email, File Transfer (FTP), Web Browsing, Data, Voice & Video applications.
- The Internet is a vast resource of information. Any information can be searched by just entering a keyword.

Who is the target Audience?
Almost everyone uses the Internet including Enterprise customers, SME, Home users, Students.

How do you access the internet?
Take a simple example. You have a mobile phone, you send something, you are connected to the nearest base station transceiver or the mobile tower. And that mobile tower is further connected to the Master Switching Centre (MSC), which is connected to the internet. So that way you access the internet.

In case of leased line Internet service at corporate offices, their traffic is sent to network

Further discussion on technologies

router in the service provider's POP. From their traffic is forwarded to the worldwide internet network.

What are the limitations of the internet?

Data transfer on internet works like your normal postal system, you write a letter, you put it in an envelope and then you drop in any letter box placed by post office. From there post office people collect all the letters, distribute them city-wise. And once your letter reaches to that city with all other letters and there, it's distributed in different areas of the city. And postman of your area delivers your letter to the recipient or addressee, to whom you have addressed the letter. He can either put it in a letter box at their home or ring the bell handover the letter to them.

In this scenario, usually your letter will reach without fail however there is no guarantee that your letter will reach. There is no assurance that your letter will not be lost.

Similar is the case with Internet, when you send a packet from your office to anywhere in the world, it's delivered usually, but there is no guarantee and assurance. However, in corporate communication when a packet is transmitted, they need guarantee and assurance that packet will reach the intended recipient. So, a reliable and secure network was required for corporate data communication. Then a technology called Virtual Private Network (VPN) was evolved. VPNs can be created over the public internet and customer could transmit their data in a secure manner with assurance and without any loss of data.

However, the service is still dependent on the public network. Since VPNs are configured on already available physical network so the reliability and availability of internet was still a concern.

What is Internet leased line connectivity?

Leased line is dedicated and always available connectivity to the customer. It's opposite to dial-up service which was available on demand and charged as per usage. Leased Line service has fixed charges irrespective of usage.

In this connectivity link from customer location to the network point of presence of the service provider is leased for that customer only. That last mile is not shared with any other customer, but once their data reach in the network router, it lands on the world wide web and travel on the shared internet Network.

Who needs Leased Line?

Corporate offices usually prefer Internet leased line because their requirement is much broader, they need higher speed, more bandwidth, so they

Further discussion on technologies

opt for lease line connectivity which is dedicated to them and always available.

Point to Point Leased Line versus Internet

In case of point-to-point link, each of the customer office had to be connected to all other offices and create a mesh Network to communicate between each other. This is complex solution and is also costly to maintain.

The Internet is a shared public network and ISP (Internet Service Provider) just extends its connectivity to the customer premises using available last mile media. So, the Internet offered an easier solution. Any company just had to get internet connection and they can communicate with all other offices who are connected with Internet.

However, the difference here is that point-to-point leased line is a secure communication between two of the customer locations while internet is like a public road where there is no guarantee of speed and there is no control on the traffic being a public network. It isn't a secure network as well. However, it provides an easier and cheaper medium of data communication for any user.

Pankaj Gupta

Multi-Protocol Label Switching (MPLS)

What is MPLS?

MPLS is a data packet forwarding technology with improved forwarding speed of routers by using labels to make data forwarding decisions.

In MPLS terminology, the packet handling nodes or routers are called Label Switched Routers (LSRs) and are classified into two broad categories:

- At the edge of the network high performance packet classifiers that can apply (and remove) the requisite labels are known as Provider Edge Routers (PE) or Label Edge Routers (LER).
- Routers that perform routing based only on Label Switching are called Label Switch Routers (LSR).

When a data packet enters the first MPLS router (PE), header analysis is done just once, and a label is attached to the data packet. Subsequent routers forward the packet by just swapping the label and so it decreases the forwarding overhead on the subsequent core routers.

Further discussion on technologies

What are the applications of MPLS?

Transfer of Data, Voice & Video for interoffice communication. It can be used to carry many kinds of traffic, including IP packets, and Ethernet.

Who is the target Audience?

Corporate customers who need a private network for corporate communication connect their multiple sites in a private, secure, and cost-effective manner with improved performance using quality of services for their data communication traffic.

How is it different than Internet service?

In a normally routed environment over the Internet, frames pass from a source to a destination on a hop-by-hop basis. Transit routers in the network evaluate each frame's Layer 3 header and perform a route table lookup to determine the next hop toward the destination. This tends to reduce throughput in a network because of the intensive CPU requirements to process each frame.

In MPLS, only provider edge router (PE) analyses the packet received from customer edge router (CE), then makes routing decision, creates a label switched path, adds its label on the packet and forwards to next router in the network i.e., Provider router (P). All other routers in the network just swap the labels till the packet reaches to destination PE which is connected to the recipient CE (customer edge router). Thus, packets are forwarded based on labels. Only labels are swapped at each hop in the core network.

Since routing decision is made only once and then packet travels on the pre-defined path, so data transfer in MPLS is faster than Internet where each router in the network had to analyse the packet and make routing decision.

Further discussion on technologies

Chapter-6

Simplifying Technical Terms

Voice Communication

Microphone

A microphone converts our audio signal (whatever we speak) into an electrical signal which can be amplified and carried over telephone lines or wireless medium to a receiver who is far away.

Speaker (or loudspeaker)

Speaker is used to reconvert the electrical signal into audio signal which we can hear as normal sound through our ears.

Speaker and microphone work similar to our mouth and ear; that is one can generate the audio signal and the other can receives and decode the audio signal. Every telephone whether landline or cellular phone essentially have a microphone and a speaker which enables two-way communication between people who are sitting far away from each other.

Frequency

Frequency is the repetition of a signal in one second (period, T). You must have studied the concept of frequency in the physics in high school.

Frequency is measured in Cycles per second or Hertz (Hz).

Frequency of a signal is calculated by the number of times a wave repeats in a second hence its reciprocal of time T, which is the amount of time taken by a wave in one repetition.

if a wave repeats two times in a second then its frequency, f= 2 Hz.

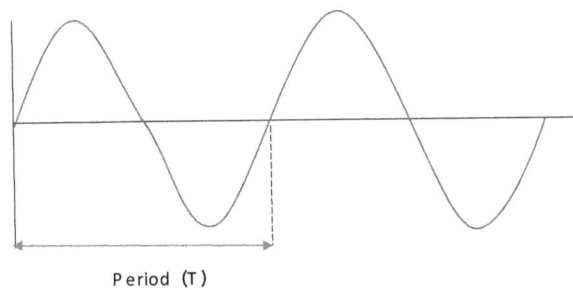

Period (T)

You must have observed the sea waves, rising, falling, and repeating the same pattern. A signal wave is somewhat similar to that in look and form.

Higher frequencies are measured in Kilohertz, Megahertz and Gigahertz viz.

- 1000 Hz=1 Kilohertz (Khz)
- 1000 Khz=1 Megahertz (MHz)
- 1000 MHz=1 Gigahertz (Ghz)

Wavelength

Wavelength is length of a wave, the distance occupied by a single cycle of a wave. In another way, it is the distance between two peaks or trough of a wave. It is represented by symbol lambda (λ).

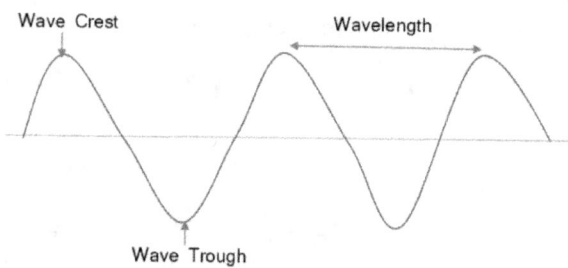

If v is speed of wave and f is the frequency of the wave, then wavelength, **λ=v/f**

Voice Frequency or audio frequency

When we speak, sound waves originated form our mouth travel in the air as waveform. Human speech has frequency components in the range from 20 Hz to 20 Khz. In telephony, the usable voice frequency band approximately ranges from 300 Hz to 3400 Hz (3.4 Khz).

The bandwidth allocated for a single voice-frequency transmission channel is usually 4 kHz, including guard bands, allowing a sampling rate of 8 kHz to be used as the basis of modulation system used for the digital PSTN.

Pankaj Gupta

Data and Signal

Data is something which has meaningful information written or recorded in any form. Signal is electric or electromagnetic form of data which is transmitted along a suitable medium to carry data from one place to another place.

Analogue and Digital Communication

Any signal which varies continuously is said to be analogue. When we speak, sound waves originated from our mouth are an example of analogue data. Another example is video image on TV screen.

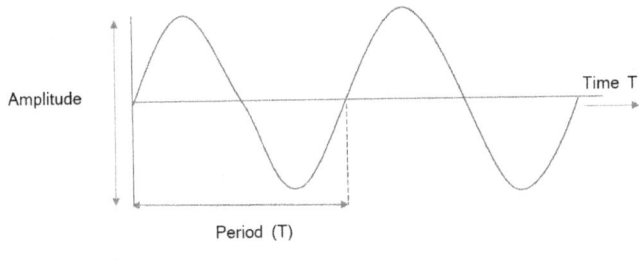

Analogue Signal

On the contrary, a signal which repeats itself after a certain period is termed as digital. It can be either 0 or 1, i.e., binary since it has only two states.

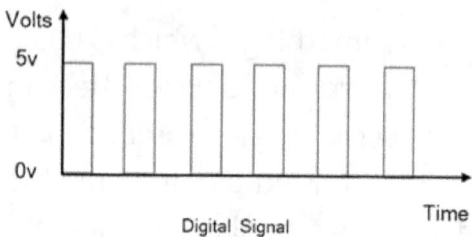

Digital Signal

In a communication system, data are carried from one point to another point by electromagnetic (EM) waves. Analog signal of continuously varying EM waves can be propagated over wire media like twisted pair, coaxial cable and fibre optic cable or wireless media such as atmosphere or space. A digital signal is actually a sequence of high and low voltage pulses which are transmitted over a wired medium.

Bandwidth and data rate

Bandwidth is the term used for defining the capacity of the communication channel. It is the width of the frequency spectrum of a signal. It's a range of frequencies constituted in a signal for transmission purpose. This is the bandwidth of a transmitted signal measured in cycles per second or Hertz.

In the transmission of a digital signal (0s and 1s), data rate is the number of bits we can transmit in a second that is "bits per second (bps)" or Bytes per second (BPS) where one byte is equal to 8 bits.

Pankaj Gupta

Higher data rates are measured in Kilobits per second, Megabits per second, Gigabits per second and Terabits per second viz 1024 bps=1 Kbps (Kilobits per second).

Bandwidth vs Speed

It's similar to the concept of physical roads where we say that there is single lane where you need to drive slowly or its two-way lane or four lane highway where multiple vehicles can drive with speed. Communication between computers is digital i.e., in form of bits (0s and 1s) and how many bits can be transmitted in a second becomes the speed of communication.

Modulation

Modulation is the process of varying one or more properties of a periodic waveform, called the carrier signal, with a separate signal called the modulation signal that typically contains information to be transmitted. Modulation can be analogue and digital.

For example, to transport a voice signal on a communication channel, either the frequency or the amplitude of the carrier wave is changed in accordance with the voice signal to carry it for the long distance.

Time Division Multiple Access (TDMA)

Time Division Multiple Access (TDMA) is a digital modulation technique used in digital cellular telephone and mobile radio communication. TDMA is one of two ways to divide the limited spectrum available over a radio frequency (RF) cellular channel. The other is known as frequency division multiple access (FDMA).

Frequency Division Multiple Access (FDMA)

FDMA allows multiple users to send data through a single communication channel, such as a coaxial cable or microwave beam, by dividing the bandwidth of the channel into separate non-overlapping frequency sub-channels and allocating each sub-channel to a separate user.

Code Division Multiple Access

CDMA is an example of multiple access, where several transmitters can send information simultaneously over a single communication channel. This allows several users to share a band of frequencies without undue interference between the users, by employing spread spectrum technology and a special coding scheme where each transmitter is assigned a code).

CDMA optimizes the use of available bandwidth as it transmits over the entire frequency range and does not limit the user's frequency range.

What is a time slot?

In time division multiple access technology, we allot the specific time periods to transmit different customer traffic. That means available time is divided in small slots and periodically made available to different customer traffic. And this period is called timeslot. A single timeslot is defined of 64 kbps. Multiples of 32 timeslots makes a carrier of 2048 kbps or 2 Mbps and also termed E1.

Time slot allocation is somewhat like musical chair game where each participant gets the chair to sit for a small period and then some other person gets the chair. That means one chair is available, but everyone will get a chance to sit for a certain period when his turn comes. In the same way timeslots are allocated to various customer to send their data traffic over a single communication medium.

Modem

An electronic device which receives digital signal from the source and convert into analogue signal to enable it over transmission channel. Another modem at the receiver end receives the analogue signal and convert back to digital signal which is to be interpreted by the receiving device.

Multiplexer

Multiplexing means many into one like mathematics where we multiply many numbers and get one number as output. In the same way, multiplexer is a communication device which receives signal from many sources (in fact that is customer traffic) and the aggregates them to be able to transmit through a communication channel either fibre medium or the satellite channel. It includes demultiplexing job as well. Multiplexer at the receiving end separate the different customer traffic and distribute to the specific channels allocated for the customers.

Cellular Technology

In cellular technology the region is divided into smaller cells which are small hexagonal areas. A cell site serves the mobile users in that cell area. Same frequency can be re-used in a different cell after a certain distance. This maximizes the spectral efficiency of the available frequency spectrum and enables provider to handle several users without interference. When a subscriber moves from one cell to another cell then call is handed off and communication goes on uninterrupted.

Pankaj Gupta

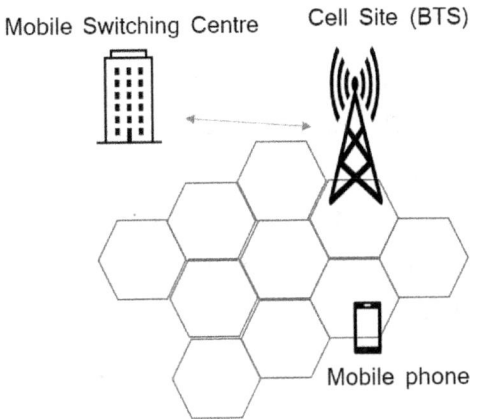

Global System for Mobile Communication (GSM)

The Global System for Mobile Communications (GSM) is a standard developed by the European Telecommunications Standards Institute (ETSI) to describe the protocols for second-generation (2G) digital cellular networks used by mobile devices such as mobile phones and tablets. It was first deployed in Finland in December 1991 but by the mid-2010s, it became a global standard for mobile communications achieving over 90% market share and operating in over 193 countries and territories.

Satellite Communication

What is satellite?
Satellite is a repeater in the sky which receives the signal from an earth station, amplify it and retransmit to another earth station in other part of the globe making communication possible between countries situated far apart physically in different parts of the world.

So how does satellite communication happen?
One transmitting station on earth transmits its signal on one frequency (termed as uplink frequency), satellite receives the signal, amplifies it, and transmit it on another frequency (called downlink frequency) to another earth station.

To make this happen first, we establish a satellite link from one earth station in India to any other earth station abroad, maintained by foreign service provider and then we test this link between two earth stations. Once this link is tested good then we allocate channels or timeslots to individual customers and provision their circuit over this satellite link as and when needed.

Pankaj Gupta

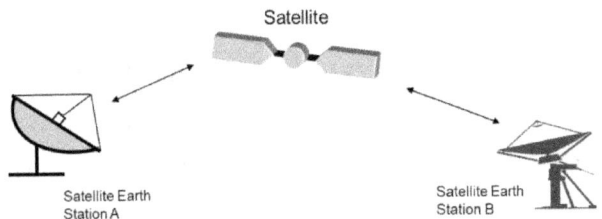

Satellite link for Point-to-Point Communication

What is the transmission equipment behind the earth station antenna?

There is a satellite modem which radiates a carrier signal on the allotted frequency and then this signal is amplified by a high-power amplifier and given as input to the dish antenna.

Bandwidth allocated to a carrier decides the capacity of the carrier that how much bandwidth it can provide to the customers. For example, one carrier of 2 Mbps has got 32 timeslots out of which 30 or 31 timeslots can be used to carry customer data traffic. 1st and/or last time slot is reserved for signalling.

Very Small Aperture Satellite (VSAT)

It provides a low-cost alternative for communication using satellite link where several small VSAT antenna can receive signal from a transmitting earth station (called Hub station) which is shared with all VSAT stations.

Small dish antenna (VSAT antenna) is installed at customer locations to communicate with hub station. VSAT can also transmit a low bandwidth carrier hence its two-way communication.

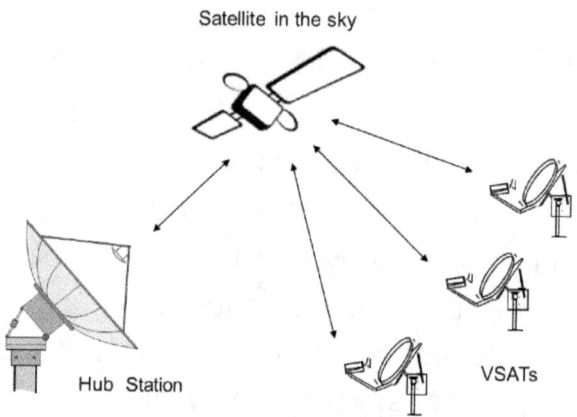

Point to multipoint link (used for broadcasting and VSAT)

Direct to Home (DTH)

You must have used this service for receiving television signal at home. This is possible due to satellite communication but in this case, you only receive from the satellite. You have a small dish installed at your home and it's connected to your "Set top box" which acts as a satellite modem in this case to convert the received signal into video signal for television.

It's not a two-way communication like other telecommunication services we discussed.

Data communication

Any information which has some message and being sent over a transport medium is termed as data communication. That means this terms data communication evolved with the evolution of computers. However, the communication over network might have started with the evolution of internet. Therefore, we assume that data communicated started with the development of inter-networks that is internet.

LAN

Local Area Network, which is a term used for computers or the network devices placed in an office or campus and interconnected with each other through network cables or wireless connectivity for the purpose of sharing data between users.

In case the computers are connected through wireless medium then it's called Wireless LAN (WLAN).

WAN

Wide Area Network, where the computer or network devices are placed far apart in different locations, may be in a city or different cities or countries but they are interconnected via some transport medium like fibre cables, microwave links or satellite links for the purpose of sharing data between offices.

Router

An internetworking device which connects two computer networks and enables communication between them.

The word router is seemed to have generated from the word 'route' that mean this is the device which decides the route. It's something like our journey in the physical world. When we start from our home and drive to our office then we take one of the available routes. How we decide it depends on the distance and time taken in travel as well as the cost of travel.

In the same way, router receives the data packet from the computers and see the destination address and then decide the best path from the available routes to send the packet on its journey over the networks.

TCP/IP

This protocol suite consists of a large collection of protocols that have been used as Internet Standards. TCP/IP models organizes the communication task into five layers.

- Physical Layer,

- Network Access Layer,
- Internet Layer,
- Transport Layer, and
- Application layer.

The **physical layer** covers the physical interface between data communication device (e.g., workstation, computer) and a transmission medium or network. The **network access layer** is concerned with access to and routing data across network for two end systems (server, workstation etc.) attached to the same network. When the two devices are attached to different networks, rules of communication needed to allow data communication between different networks. This is the function of the **Internet Layer** and **Internet Protocol** (IP) is used at this layer to provide the routing function across multiple networks.

Transport layer is concerned with the reliability of the communication that assures that the data transmitted reaches the destination in the same order. Transmission Control Protocol (TCP) is most commonly used protocol for this function. **Application layer** contains the logic needed to support the various user applications e.g., file transfer etc.

Internet Protocol (IP)

Protocol is rules of communication. Internet Protocol (IP) is an internetworking protocol used on the Internet. It is part of TCP/IP protocol suite and interfaces with higher layer TCP during communication.

IP Address

It's a unique identification of a device on the internet. It's like the physical address of your home which includes unique house number, colony name, city, and pin code. Similarly, any computer, router or network device which is part of the internet is assigned a unique number for its identification on the network.

An IP address contains a 32-bit global internet address consisting of a network identifier and host identifier in IP version 4 (IPv4) while its 128 bits in length in IPv6.

DCE (Data Communication Equipment)

These are electronic devices which initiates data transmission to the network. Modems and Multiplexers are example of DCE shown in the network diagram of point-to-point leased line (Fig.3, page no. 23).

DTE (Data Terminal Equipment)

These are terminal devices which is the end equipment connected to the network through

DCE. For example, Routers act as DTE when connected to the modem (DCE) in the network diagrams of point-to-point link (Fig.3, page no. 23).

Transport media.
Like the physical world where we use different means of transport to travel for long distance, in the network also we need transport mediums to send the data through the network for long distance communication. Copper wires, fibre cables, Wireless Radios, and Satellite are examples of the transport medium typically deployed in the communication networks.

Latency, Round trip delay (RTD)
Like we measure the travel time in the physical world, latency is the travel time for a data packet from source to destination. Round trip delay is the time taken for a packet from source to destination and receiving reply at the source. You can compare it with your travel time from home to office and then coming back to home.

Ring Topology
When fibre is laid for the telecom networks, its usually installed in the ring form. That means from source to destination, there will be two fibres using different routes. So, if one of the fibres got cut or failed due to some reason then data traffic will be routed through the other fibre without

significant interruption in the data communication.

Self-healing

When one arm of the fibre ring got cut or damaged then other arm of the ring will take over the traffic automatically. This is termed self-healing.

Undersea cables

Fibre cable laid through the sea for the communication purpose between different countries.

Fun time

Packet forwarding mechanism in MPLS.

Here is an interesting demonstration how packet is forwarded in MPLS with the help of famous Bollywood characters and their most famous dialogues:

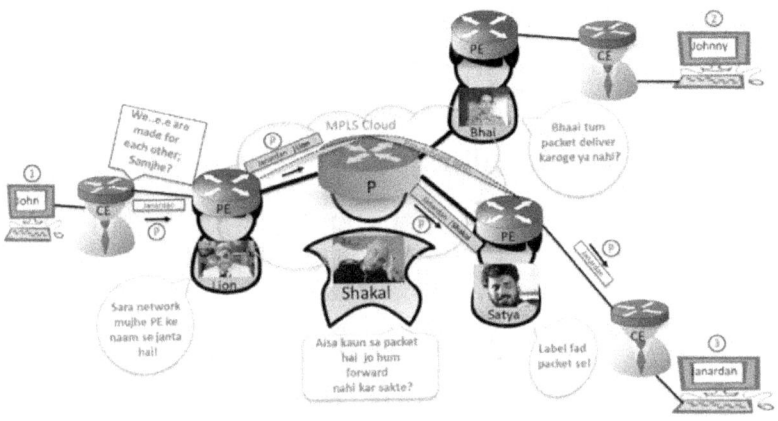

There are three customers (John, Johnny & Janardan) who are connected through their Customer Edge Routers (CE) which are further connected to Provider Edge Routers (PEs) in the MPLS network through any last mile. A CE can send packets to PE only.

Now let's see how a packet is forwarded in MPLS network from one location to other location.

- If John wants to send a packet (P) to Janardan, CE at John's office sends the packet to connected PE (Lion).
- PE (Lion) is connected to the Provider router (Shakal) in the core MPLS network. PE(Lion) analyses the packet's destination, decide the route, adds its own label to the packet and forward to P(Shakal).
- P(Shakal) swaps the labels i.e., removes Lion's label and add its own label and forward the packet to next PE router (Satya) on pre-defined path.
- PE(Satya) removes the label from packet and delivers the original packet to CE at Janardan's office.
- If the packet is destined for CE(Johny) then P(Shakal) forwards the packet to PE(Bhai) who delivers it to Johny following the same label swapping process.

Afterword

Hope you have had an interesting journey and enjoyed travelling through my story at different stages of developments. Technology is a vast subject, and it may not be covered completely in a small book. Moreover, technology has been evolving and changing continuously. What we see today, might become obsolete tomorrow.

The aim of writing this book is to share my knowledge and experience which I have gained during decades of corporate journey to give you an overview of the telecom world and help you to learn the basics in a simplified way. Some standard definitions and universal facts are taken from public resources on Internet for references and are not my copyright.

I have placed the definition of technical terms in later part of the book so as not to interrupt the flow of the story as I have used the story narration format to pass through the milestones of my journey and learning during phases of development.

For further reading or training on a specific topic you may email me or message me on my social media pages given at the end.

~ 84 ~

Cloud Services

Simplifying Cloud Computing for Everyone

DEMYSTIFYING BASICS

Pankaj Gupta

Chapter-1

The Concept

What is Cloud?

You all have heard of clouds. Cloud is formed of water droplets; the cloud is aggregation of water droplets. So, when cloud rains it drenches everything in the area below it. That means it's a common resource which is available to people in shared environment, although it's a natural resource. Now the same concept is replicated in cloud computing business model where shared resources are made available to everyone. You just need to access it.

Let me simplify it in terms of real-life utilities. We all have been using Electricity services at our home and offices for decades. How do we get electricity connection?

To subscribe, initially we pay one-time installation charge for the minimum hardware required at our premises i.e., wiring and meter. Then we have to be wired to the nearest transformer installed on an electrical pole which is further connected to main powerhouse. Thereafter we only pay for the minimum subscription charge (monthly rental) and the usage charges.

Similar is true for water supply services also. In olden days we used to dig well in our homes or at a common place in the village. Then we had to fetch water from the water well and stored in buckets. With development in civil engineering, we constructed water tanks in our colonies and created a common storage of water which became a shared resource for everyone in that area.

Simplifying Cloud Computing

Here also we just need to install minimum hardware required at our premises and get connected to the nearest water tank through water pipes then we start getting continuous water supply. Further, we only have to pay a nominal amount as monthly rental and bill as per usage.

Cloud service model is analogous to these traditional electricity and water services where we minimise the hardware required at customer premises and centralize the utility resources at a common place usually called "Data Centre". This leads to migration from "CAPEX business model" to "OPEX model" i.e., reducing the infrastructure cost (Capital Expenditure-CAPEX) for users and making shared resources available at nominal recurring cost (Operational Expenditure-OPEX).

Cloud Business Model

You all must be using various applications like emails, office outlook 365, CRM etc., on daily basis. Well, the concept is similar. You have a hardware device at your home or office (Mobile phone, Laptop or Desktop Computer) and a wireline or wireless connection for accessing the servers installed at any service provider's premises i.e., data centres in cloud environment.

This way you get access to storage space, files, application softwares and databases as per your need. Thus, it is on-demand and pay per use service on contrary to traditional investment in IT infrastructure and fixed operation cost.

Pankaj Gupta

Internet

We all have been using web browsing and web-based email services through Internet for more than a decade now e.g., outlook email, Gmail, yahoo email, are a few but there may be many more. But what is Internet actually?

Internet is a network of networks. Now what is a network?

Network means added or connected devices like computers, servers, or any other element. Internet started with connection of four computers in a lab for a single organization but expanded further into a global network i.e., it is a collection of computers spread across globe but interconnected with any of the transport mediums (Copper, Fibre, or Satellite).

To make it easy to understand, let me put this way "Computer network is similar to our people's network connected through any social media platform. And networking means adding computers like we add people in our social network."

Suppose you have a family of four person in a house connected with each other and further you are connected with three other families of four persons. So, this network of families can keep expanding. In the same way Internet has been evolved and it has become network of networks connecting millions of elements now.

Internet is also represented as cloud since it is formed of millions of computers and shared by people globally. In a simple way, it is depicted in the figure below.

Figure-1: A representation of Internet

Explaining Cloud Computing

Let me take the example of email services again. We have been using the webmail services (e.g., yahoo mail, Gmail etc.,) for personal use as well as for small businesses. These web email services are very good example of cloud services where we just need to connect to email server on internet, then login using our user's name and password and then we can use the email services; reading mails, sending, and receiving mails etc.

In past, enterprises had to set up their own mail exchange server and maintain it and have their own control. Traditionally enterprises had been investing in creating IT infrastructure at their premises and maintained it. They usually identified a separate room to install IT equipment and they called it Server room or Computer Room or Communication room as per prevalent practice, available technology, and privacy. For all these practices an enterprise had to invest in their own infrastructure and hire the IT resources to maintain and manage it, so this is CAPEX model.

For example, In late 1990s I worked in an organization where we created our own infrastructure of email server, domain name server and web server. So, we spent a huge amount in purchasing, installation, and commissioning of these elements. This was an example of Capital Expenditure.

Development of Data Centres provided an option of Co-location services where enterprise could rent power and space in the service providers premises and created their own network infrastructure off-site in data centres and accessed remotely through internet.

"Now with the evolution of cloud concept, service providers have created a common infrastructure of email servers, domain name servers, web servers, storage and databases where they invested in these elements and any organization needed theses services can subscribe and share the same infrastructure."

So, now enterprises and users only need to subscribe to ready-to-use infrastructure and pay only for the subscription and usage charges as per their usage hours or whatever is the billing unit. Thus, users have to bear only operational expenses. This is OPEX model.

Thus, **Cloud Computing offers on demand services for computing, storage, software, and other applications with usage-based billing.** Using services in cloud is essentially migration from CAPEX to OPEX model.

Now, what does a user need to reach cloud?

An access medium (like wireless, copper, or fibre cables) to reach the shared infrastructure in cloud. Then using their mobile phone, laptop, or desktop and through internet they can reach the IT infrastructure in the cloud.

A simple demonstration of the cloud services model is shown in the figure below.

Simplifying Cloud Computing

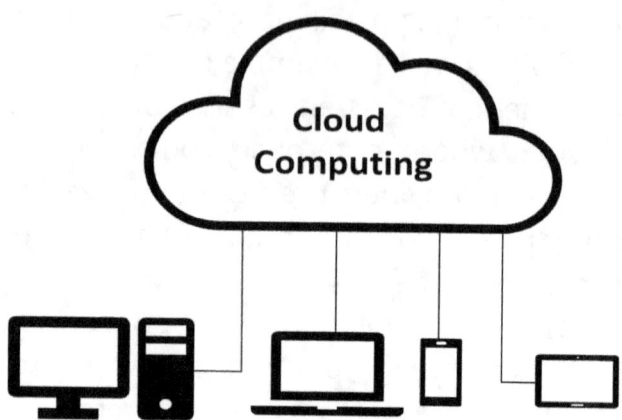

Figure-2: Cloud Computing Concept

Wish we have an access medium to reach the clouds in the sky so that we could make them to rain whenever we needed.

Advantages of Cloud services

1. Small businesses don't have to purchase so many computers and servers to make their IT-infrastructure as they can subscribe to already existing infrastructure in cloud and pay as per usage.
2. Individual users also don't have to buy hard disks to store their data, they can subscribe to any service provider and use storage space from their shared

infrastructure in cloud. Best example is Google Drive storage space which we get with Gmail.

3. Enterprises can subscribe to emails and web hosting without creating their inhouse email server and web server.
4. Only minimal hardware and software required at customer's premises i.e., laptop, desktop, tablets or mobile phones so very low expenses and they only have to pay for a nominal subscription charge and/or usage charge.
5. Upgrades and downgrades are easy as you just have to place order with service providers instead of buying additional hardware and software in CAPEX model.
6. Service commissioning is faster as all the resources are already in place with service providers. They only need to add you as user and grant access to the required resources.
7. Users don't have to employ IT-support staff to maintain their in-house infrastructure like CAPEX model. A common resource of IT-staff at Data Centre maintain the IT-Infrastructure and they can also provide remote support for users in need.

Chapter-2

Cloud Computing Services

What are the types of cloud computing services or business models?

Well, there are different ways we can use the cloud computing services.

- Infrastructure as a Service (IaaS)
- Platform as a Service (PaaS)
- Software as a Service (SaaS)

Let's know a little bit about these options.

Infrastructure as a Service (IaaS)

It can provide you whatever computing infrastructure you need in cloud environment like physical hardware with operating systems, virtual servers, storage space, and database connections. It can also provide you separate dedicated hardware, software, and storage as per your need.

This service gives you flexibility and management control over your IT resources although physically hosted in service provider's cloud. You don't have to purchase the resources and bother for installation and maintenance.

For example, Amazon WorkSpaces provides virtual desktops in the cloud for End user computing. Google Drive is another example of IaaS where you can store your files in the cloud and share with others.

Platform as a Service (PaaS)

It provides tools and platform for software developers and application programmers in cloud environment for creation of application, webpages, and other softwares. For example, if you want to develop your own website then you can subscribe WordPress services to use their website builder and design tools for website development.

This service helps you to focus on management of your applications only as you don't have to worry for procurement of hardware, software and updating patches. For example, Amazon Developer Tools provides software development service on AWS.

Software as a Service (SaaS)

Any software application you access on Internet and start using is like SaaS. Web-based emails, MS-Office 365, Salesforce are few examples. So, instead of buying the software you only subscribe and pay for use.

These are ready to use applications or software from user perspective. Users do not have to download and install email client on their device. They do not have to maintain the software application. They just have to use internet browser to access the email service.

For example, I am using Gmail service which I can open anywhere on any device; mobile phone, laptop, or desktop. I just need internet connectivity to access the Gmail server. Google Docs is another example where you can create word document and share with others for their inputs and modification.

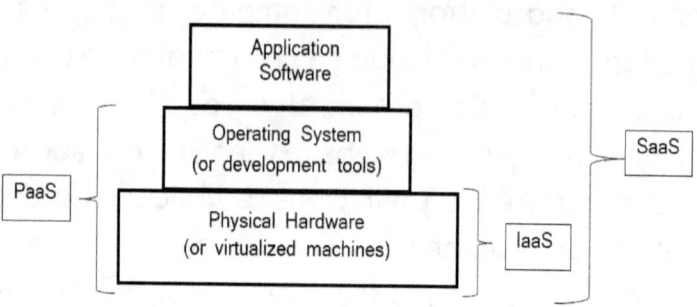

Figure-3: Explaining Cloud Services

Figure above shows a simple representation of the cloud service models. In IaaS, you only use infrastructure, in PaaS you use infrastructure and development tools and in SaaS you use application software in addition to previous two layers. But as a user you don't have to worry about the underlying hardware and operating systems.

Chapter-3

Cloud Deployment Models

Various cloud service delivery models only differ by their deployment strategies; how the cloud infrastructure is deployed, though the concept of the cloud will remain same.

Let's understand in a simple way. Different cloud models are defined by physical boundaries like you have a small housing society made of few hundred houses and there is a pool of common facilities like club, sports ground, play school etc., are available to local residents. Then you have a bigger society made of thousands of houses with pool of common resources available to share by society people. And then you have a city made of so many societies and colonies with pool of common resources available to all public on subscription basis.

It is somewhat similar to network topologies like Local Area Network (LAN) which belongs to a single organization, Metropolitan Area Network (MAN) which connects a city and Wide Area Network (WAN) which spreads globally. Likewise, we have Private Cloud, Community Cloud and Public Cloud. Let's understand a bit about them.

Private Cloud (On-premises)

In this model, IT infrastructure is made for a single organization and used for inter-office business transaction's purpose. Here the computing resources are owned and operated by a single organization and not shared by other customers. It can be at the customer premises or in the data centre. So, it's almost similar to the traditional IT infrastructure but with the benefits of cloud computing features.

Figure below depicts the concept of private cloud.

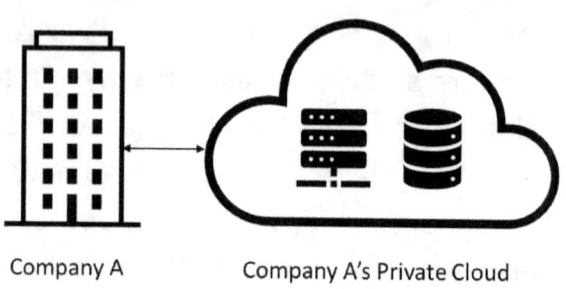

Figure-4: Private Cloud

Private cloud is preferred by the companies who need confidentiality and deal with sensitive data or to meet their regulatory compliance. It may be compared to your private locker in a bank where you only have the key and access the same though its physically located in bank.

Community Cloud

In this model, the computing resources are made available to a specific community or a group of organizations. A shared IT-infrastructure is set up in data centres in cloud environment for a group of companies with similar interests like a chain of hospitals or offices of a banks or financial institutions or any other business houses.

Public Cloud

Here the IT-infrastructure is made available to all public and deployed by government, educational or a business organization. Cloud Service Providers (CSP) own and maintain the infrastructure and access to the resources is offered to multiple customers on subscription basis or pay per use billing. Cloud provider handles all the privacy, and security issues for all subscribers.

Figure below shows a public cloud shared by many companies.

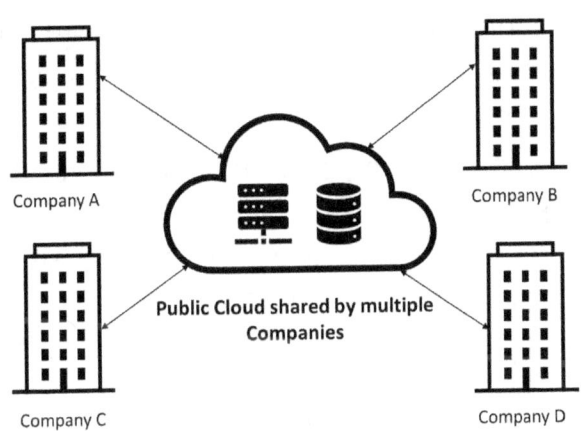

Figure-5: Public Cloud Model

Just to share that there is also a fourth option of "Hybrid Cloud" which is a combination of two or more cloud model which I am not discussing here to maintain simplicity.

Chapter-4

Related Terms

Data Centres

Data Centres are essentially part of cloud; a cloud may be formed by one or more data centres connected through any transport medium like national and long-distance fibre links, International submarine cables or satellite links.

Data centres may have physical computers, servers, storage devices like hard-disk system with security mechanism to protect from unauthorised access virtually as well as physically, i.e., data centre facilities are in highly protected area with multiple level of security checks in place while entering the building premises to allow physical access to authorized persons only.

A partial view of data centre facility is shown below.

Figure-6: Data Centre

Virtualization

With development of virtualization technology, it is now possible to create multiple Virtual Machines (VMs) and virtual server on a single physical machine hardware. Each virtual machine works like an independent computer system or server with its own operating system and application softwares though hosted on a single physical system.

Required resources are allocated to virtual machines from the physical system through a software called hypervisor.

Figure below shows a simple representation of virtualization concept where multiple Virtual Machines (VMs) are created on a single physical system.

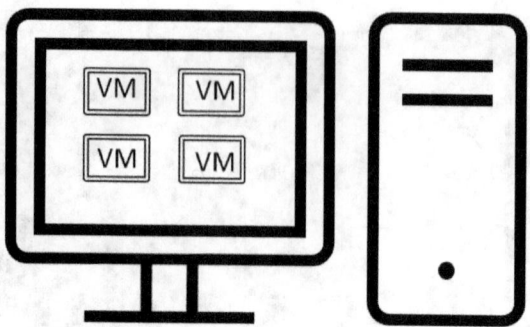

Figure-7: Virtualization Concept

Virtualization reduces the need of physical hardware and it is quick to create the resources and upgrade/downgrade as per customer needs. It also reduces the cost of IT infrastructure for service providers as well as running cost for customers.

Internet of Things (IoT)

IoT term refers to the aggregated network of collective devices spread across locations (home, business, or offices etc.,) which communicate with the help of technology and facilitate remote monitoring and management. Things to be monitored are embedded with sensors and software for connecting through internet to communicate their status. All the end devices are connected to Wi-Fi or a secured Local Area Network (LAN).

IoT extends the internet connectivity to variety of devices at home and offices which can be monitored and controlled remotely thus reducing the human efforts and cost of in person monitoring.

Figure below shows a representation of IoT model.

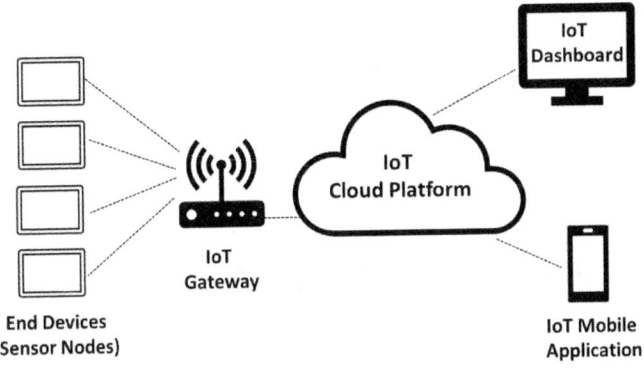

Figure-8: IOT Model

Official Definition of Cloud Computing

Let's become a bit technical for a moment. As per National Institute of Standards and Technology's (NIST), here is an official definition of Cloud Computing:

"Cloud computing is a model for enabling ubiquitous, convenient, on-demand network access to a shared pool of configurable computing resources (e.g., networks, servers, storage, applications and services) that can be rapidly provisioned and released with minimal management effort or service provider interaction."

The NIST definition lists five essential characteristics of cloud computing: on-demand self-service, broad network access, resource pooling, rapid elasticity or expansion, and measured service. It also lists three "service models" (software, platform, and infrastructure), and four

"deployment models" (private, community, public and hybrid) that together categorize ways to deliver cloud services.
Reference:
https://www.nist.gov/news-events/news/2011/10/final-version-nist-cloud-computing-definition-published

Future of cloud computing

In the past decade we have seen continuous migration of services from own infrastructure to cloud environment. Cloud services will continue to grow in future and give many benefits as it is flexible, scalable, and cost effective. In addition, it will create new application and services which may lead to creation of new job roles. To summarize, cloud computing will transform the ways we work and do businesses.

Popular Cloud Services

Amazon Web Service (https://aws.amazon.com)
Microsoft Azure (https://azure.microsoft.com)
Google Cloud Platform (https://cloud.google.com)

Pankaj Gupta

From the Author

This section is added to give you a basic understanding of the cloud computing services and business model. Hope I am successful in my endeavour and you are able to get the concept.

Your views and feedback may be communicated at the given email or message me on my social media pages:

Email address: crossthebridgewithme@gmail.com
Instagram Id: Cross_the_bridge_with_me
LinkedIn URL:
https://www.linkedin.com/in/pankaj-gupta-0188135/

Related Reading

https://www.amazon.in/dp/B09QL5ZVSX

Pankaj Gupta

https://www.amazon.in/dp/B0C65CL2V8

www.ingramcontent.com/pod-product-compliance
Lightning Source LLC
Chambersburg PA
CBHW070245220526
45465CB00004B/1536